To Mike

Peace

Mike [illegible]

Michael Greenlar
www.mikegreenlar.com
chachaway@aol.com

Thank You
Linda Bogart-Greenlar and Marissa Greenlar;
John Mannion, Jeffrey Hoone, Hannah Frieser, and the
staff at Light Work; Doug Eaton; Ellen Marie Hannan, Renée Gadoua

Pre-Press
Light Work
www.lightwork.org

Printing
Eastwood Litho, Inc.
www.eastwoodlitho.com

Layout
WRKDesigns
www.wrkdesigns.com

EMH Graphic Design

ISBN: 978-0-615-47704-6

REMNANTS
OF A SECRET WAR

MICHAEL GREENLAR

Rice paddy in Phonsavan.

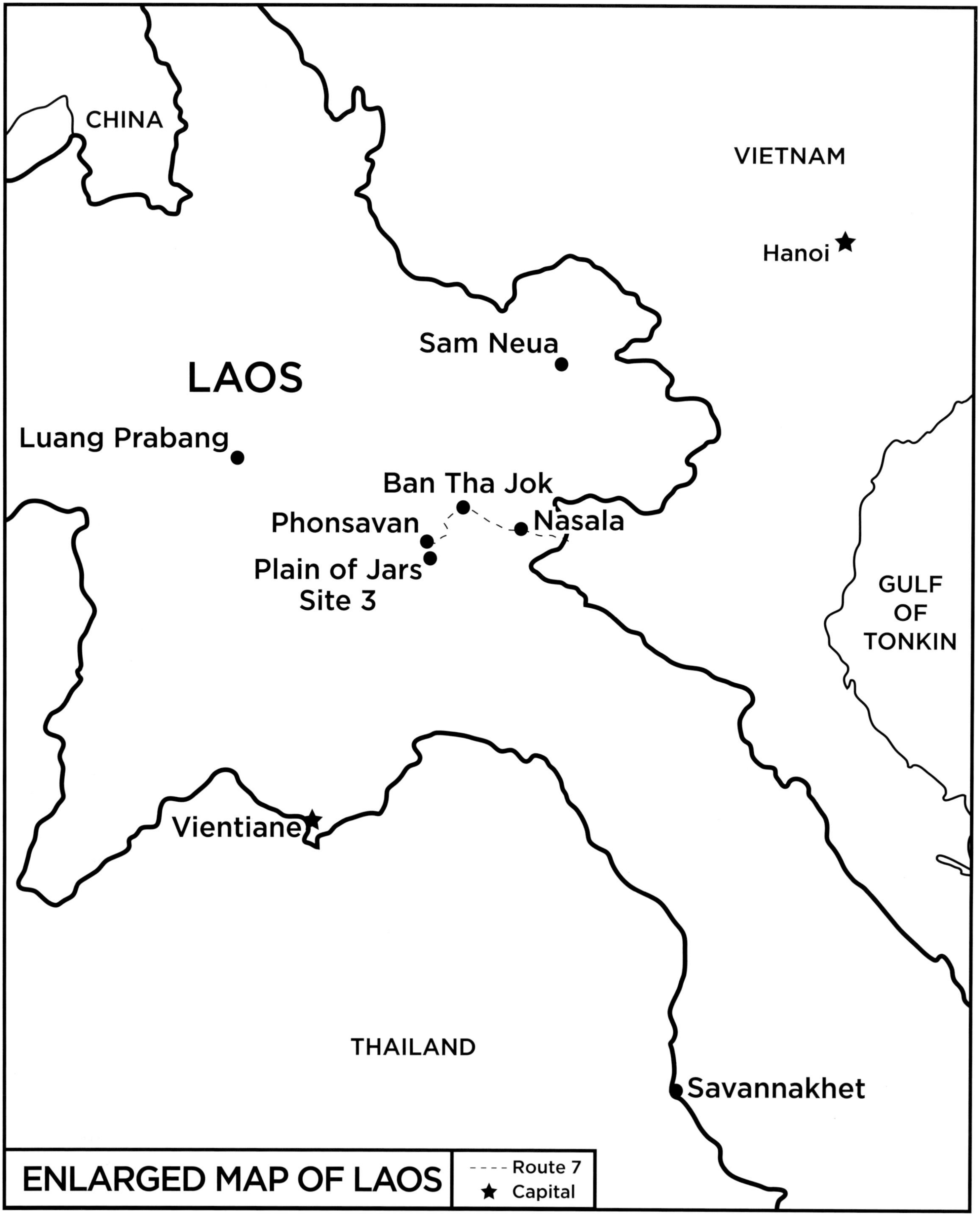

ENLARGED MAP OF LAOS

REMNANTS OF A SECRET WAR

Metal strikes metal. It shatters the quiet of the remote Hmong village of Nasala in northeast Laos. His anvil made of an artillery shell, Chong Toua Yang forges tools from scavenged Vietnam-era war scrap. These are used by the people in his village for subsistence farming. With no natural source of metal, Hmong blacksmiths have "beaten swords to plowshares" since the 1970s. Such resourcefulness and resiliency have helped the Hmong survive in postwar Laos.

The story of the Hmong in Laos is a story of survival. Driven from China in the 1850s, the Hmong settled in the isolated mountains of northeastern Laos, near the famous Plain of Jars archeological sites. After World War II, these nomadic slash-and-burn rice farmers were recruited, first by the French and then by the Americans, to be the front line of defense against the North Vietnamese army.

Because Hmong territory bordered North Vietnam and was near the start of the southern passage for the Ho Chi Ming Trail, it became a fierce battleground in the Vietnam War. This territory became the western corridor for Operation Barrel Roll, a U.S. covert bombing campaign from 1964 to 1973. This bombing campaign, along with Operation Steel Tiger in the southern region, gave Laos the distinction of being the most bombed country in the history of warfare. To provide a sense of scale, on one trip my guide showed what he claimed to be an official U.S. list of missing in action and killed in action pilots over Laos during the war. It was printed on a waxed fax roll over 30 feet long.

During those years of heavy bombing, people in Xieng Khoung province lived in caves for protection. Schools were set up in caves, and farmers tended their crops at night.
Feelings run deep about those years. To the Pathet Lao communists, the Hmong were hill-tribe mercenaries. The contemporary Hmong rebel army still fighting the Lao army in the same mountainous regions call themselves U.S. war veterans.

Both Hmong and Lao continue to farm this land despite a 10 percent to 30 percent rate of unexploded delayed-action ordnance, mostly cluster bombs. In such an environment, burning fields, planting and harvesting rice can be high-risk activities.

Ten Trips to Laos
The inspiration for this story came in 2000, when I heard Anne Garrels' report on NPR marking the 25th anniversary of the end of the Vietnam War. Within months, I was on my way to Laos. The very first descent into the Xieng Khoung Airport in Phonsavan provided a striking scene, a pockmarked hillside that resembled large sand traps on a golf course.

During that first descent, I also chose my guide, Sousath Phestry, whom I came to depend on for each of my 10 trips, from my tour book just before touching down. Sousath speaks numerous languages and claimed to have the best restaurant in town. His guesthouse was decorated with a variety of defused cluster bombs. He prided himself as an Indochina war guide but earned most of his income shuttling tourists to the numerous sites on the Plain of Jars. It seemed clear that I was off to a good start.

All of the photographs in Remnants of a Secret War were taken in Xieng Khoung Province between 2000 and 2009. The Hmong were photographed in two resettlement villages: Nasala and Ban Tha Jok.

On my first trip, a representative of an American NGO in Vientiane, the capital of Laos, brought me to a hospital and showed me prosthetics manufacturing. I was led into a small room with paint peeling from its walls and with rusty metal beds. In one bed lay a child who had undergone several operations on his arm to remove ball-bearing-like shrapnel. The doctor had given the boy's father a piece of metal that had been removed. A grim souvenir.

Later that day, I was brought into a nearly empty hospital room where what looked like a 10-year-old boy lay twisted on a bed. I was told he had been injured by a cluster bomb explosion. His head was partially covered with a blanket. I saw some scarring near his forehead. His only movement was winding a clock whose loud alarm sounded every 10 to 15 seconds. An anxious father sat bedside.

During my travels, many people told horrifically similar stories of children left alone during the growing season while parents tended the rice fields. In the story, a child about age 8 finds a metal ball. He hides it until the next day when he can share it with his friends. Then, with a circle of children sitting around, the child strikes the metal ball in an effort to open it.

The metal ball is a cluster bomb. When it explodes, it hurls ball-bearing-like shrapnel in every direction.

The pervasiveness of cluster bomb munitions became evident on my first trip. The tennis-ball-sized bombs seemed ubiquitous, with everyone talking about them, everyone affected by them. They called them "bombies."

At the conclusion of my first trip to Laos, I had photographed the following: Lao soldiers de-mining a road, a Hmong funeral, cluster bomb victims, and a blacksmith forging grass rakes from old artillery shells. I had shot over 40 rolls of 35mm film, both color and black and white. Back home, several days later, I discovered that a camera malfunction and a processing error had ruined almost all of my work. Three weeks later, I was on a plane for my second trip.

On one of my early trips, I befriended a Hmong shaman. I photographed his family repeatedly, especially his mother who when I first met her had just returned from living with Hmong rebels in the forest. She looked frail and emaciated. It was the Hmong New Year, just after the rice harvest. She wore an amazing number of strings tied around her wrist, symbols of goodwill sentiments to preserve the soul and to ward off evil spirits.

Over the years, I photographed close to a dozen funerals in the two villages. For these Hmong funerals, cows were killed to trade the cow's soul with the devil instead of the deceased's soul. I joined the men for a shot of homemade rice whiskey, sometimes drunk from a bamboo stalk.

Gilded paper that lay on the deceased during the ceremony was later burned to enrich the soul of the recently departed.

The customs, rituals and ceremonies of the Hmong were always most prevalent and predictable at the New Year, after the annual rice harvest, usually in late November or early December. The occasion is marked with village bull and water buffalo fights, feasts and special ceremonies.

One ritual I witnessed at the end of the New Year was the Free Spirit ceremony. The ceremony commemorates all who died in the village in the past year. A small table is set up with a picture of each of the deceased along with a lit candle. All of these altar items rest on a rice sifter. There are three days of crying, eating and paper burning. On the last day, the items, along with the rice sifter, are dissembled. The drinking of rice whiskey concludes the ceremony.

From blacksmithing to rituals, I saw a lot in my 10 trips to Xieng Khouang Province. The legacy of the "secret" bombings of Laos from 1964 to 1973 takes many forms. In this collection of photographs, I share some of them.

Before Remnants

I grew up in the shadow of my father, a decorated veteran of the Battle of the Bulge. Like a lot of kids of my generation, I played "army" a lot. Little did I know that a real war raged inside my father. After years of counseling and electroshock, he ended his struggle. I found his body when I returned from school one day in March 1965, the same month the United States committed ground forces in South Vietnam. I did not understand it then, but as one war had ended, another had begun.

In 1985, on an assignment for the Syracuse Herald-Journal, I traveled to Washington, D.C., with two Greyhound buses of local Vietnam veterans. The story was about the veterans' first look at the Vietnam Veterans Memorial.

It was past midnight when we pulled into the mall. The black stone of the memorial melded with the night sky. Sobbing and wailing began to break the stillness. That night, I just listened.

The following afternoon, the men revisited the memorial. My best photograph that day was a veteran kneeling in front of the name of a friend who didn't make it home. That veteran told me later that he had second thoughts about giving me his name; he had never told his mother that he had been deployed to Vietnam.

While waiting to board the bus back home, the battle-tested veterans looked across the parking lot where young sailors in dress whites posed for their girlfriends' snapshots.

Michael Greenlar

REMNANTS OF WAR

Defused bombs on Route 7 in the city of Phonsavan.

Lao soldier looks for unexploded ordnance along a road.

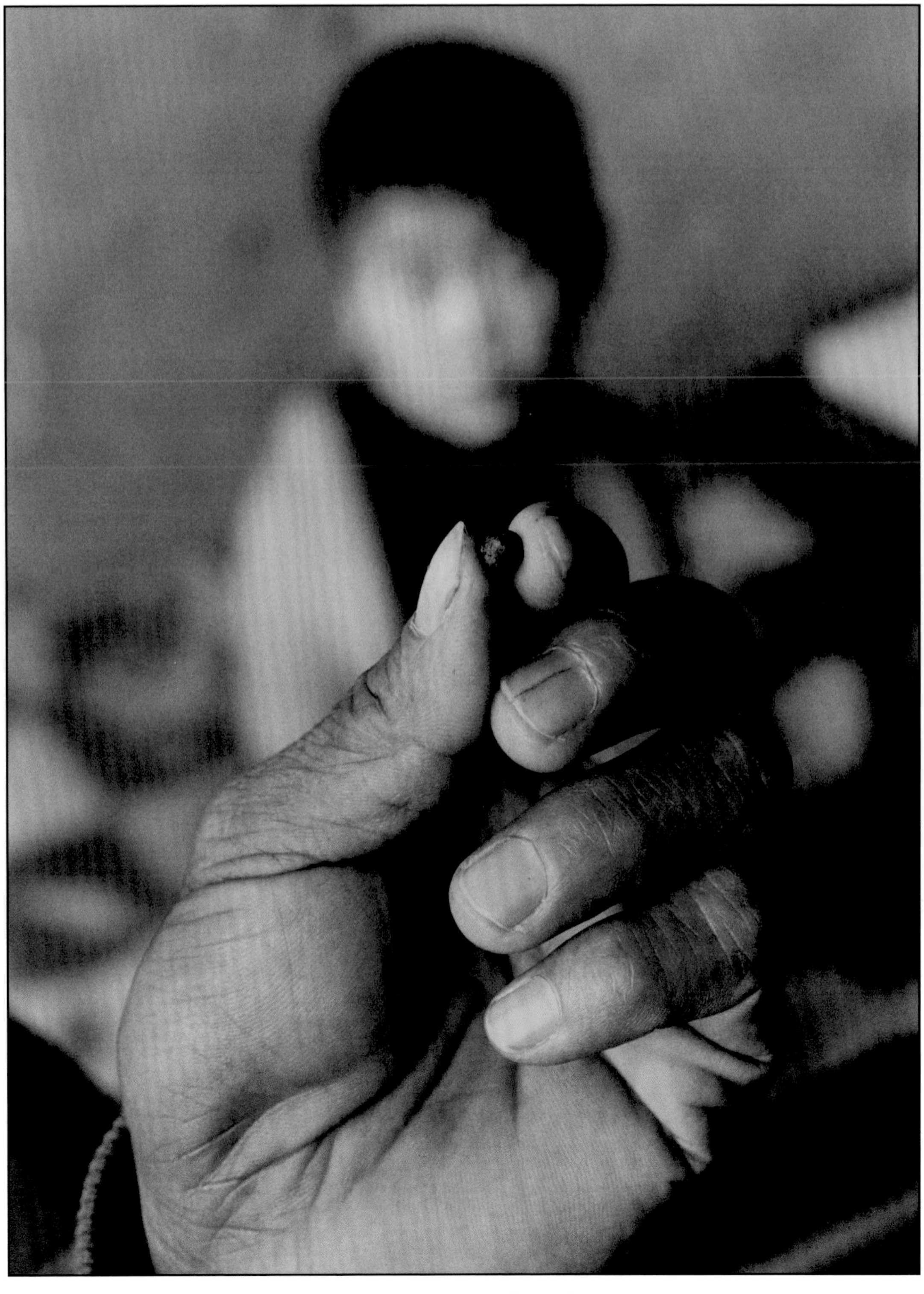

Father holds cluster bomb shrapnel removed from his son's arm.

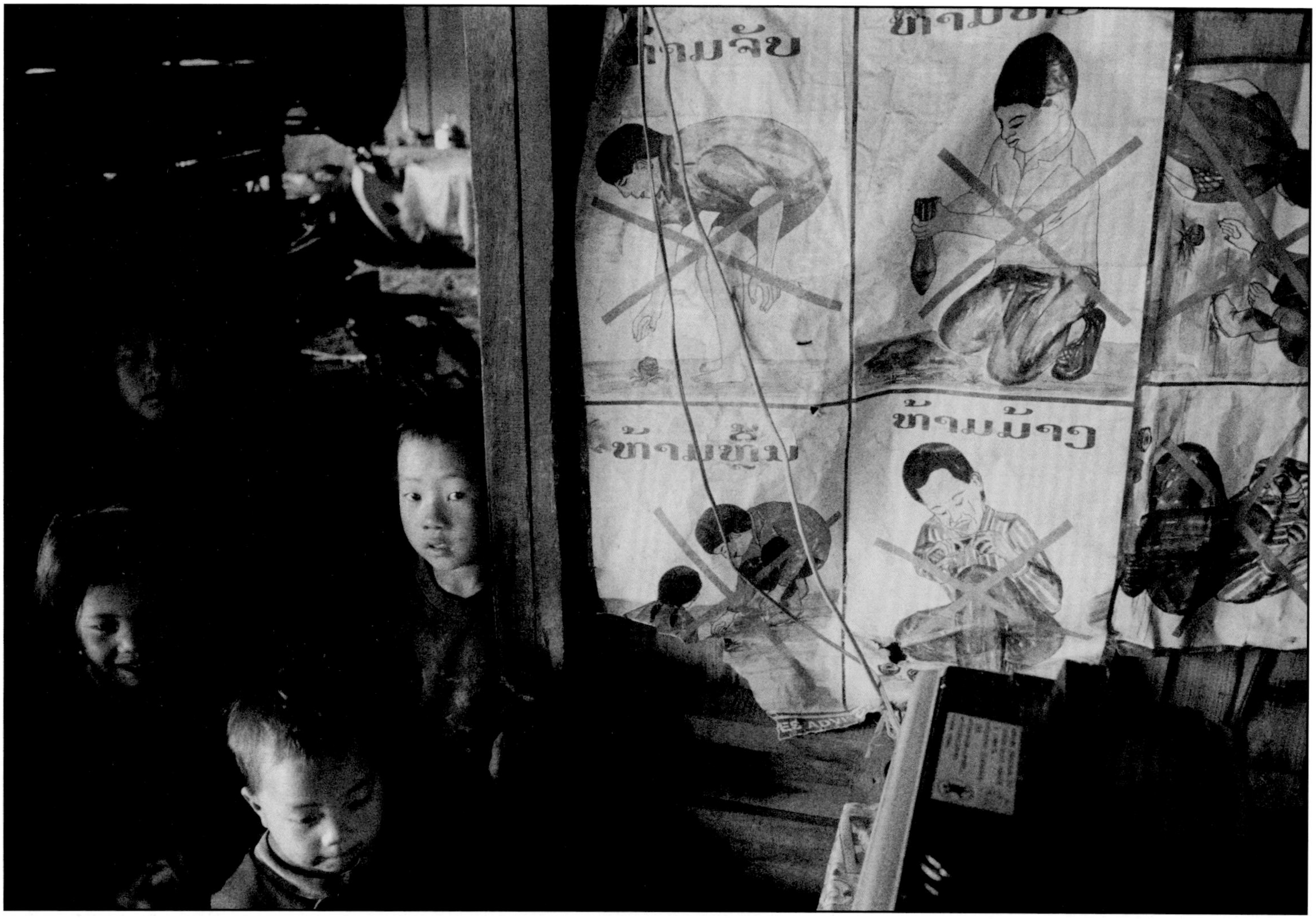

Posters warn children of dangers from unexploded ordnance.

Children gather around cluster bomb casing.

A horse eats from a cluster bomb casing.

Spring onions grow inside a cluster bomb casing.

Children peer through a garden fence made of cluster bomb casings.

A double amputee cluster bomb victim with daughter.

A former soldier holds drawing depicting bombing.

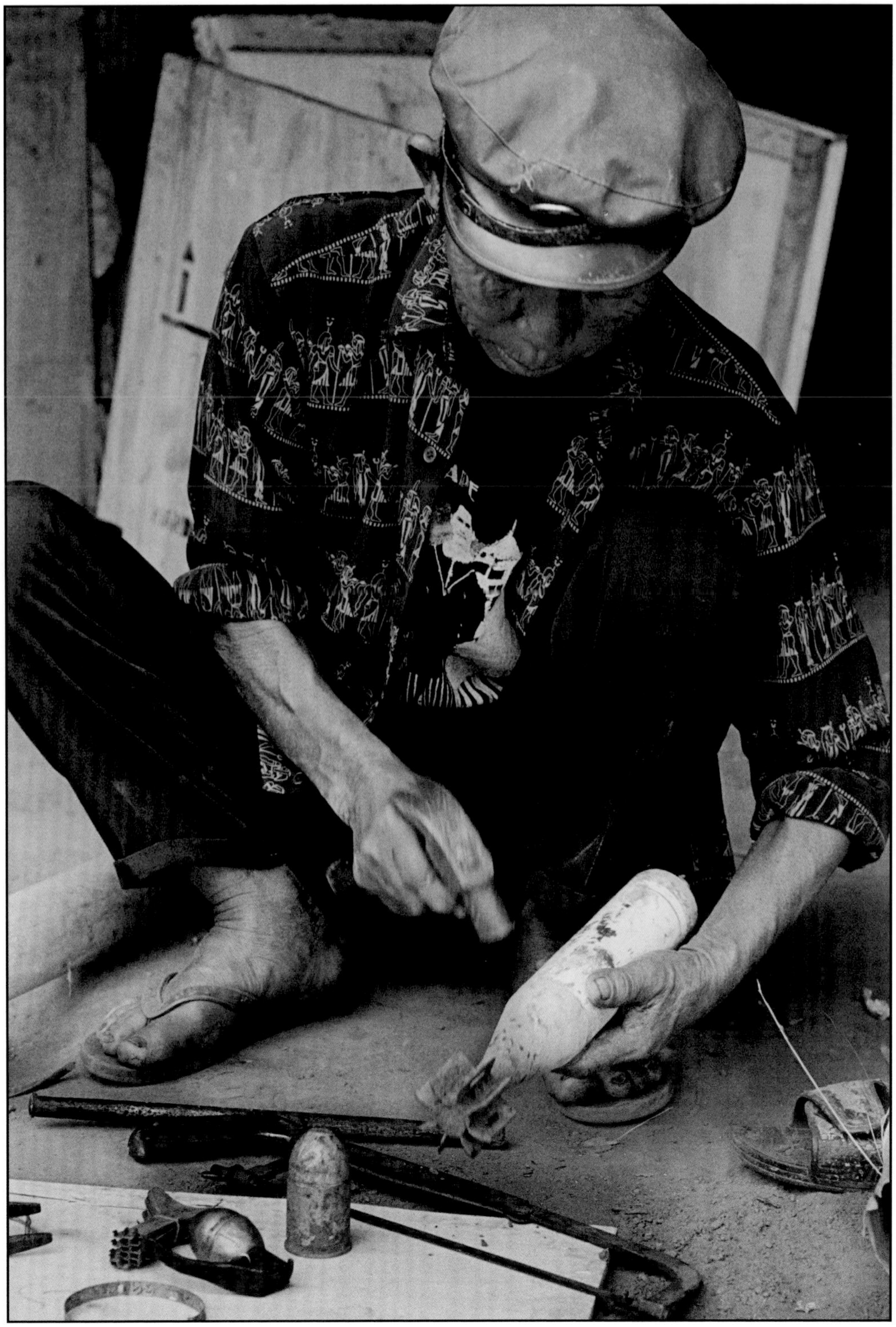

He taps a live mortar shell, extracting gunpowder for sale.

A grass rake is forged from a bomb, using an artillery shell anvil.

Grass rakes forged from defused bombs.

A blacksmith washes his hands after forging grass rakes.

Spoons made from melted aluminum shell casings.

Former defuser holds Chinese rockets, as he sits with American bombs.

A scrap yard worker greets her children at day's end.

A Lao soldier holds a newly unearthed cluster bomb.

LEGACY STATS

From 1964 to 1973, the U.S. dropped over 2 million tons of ordnance over Laos in 580,000 bombing missions, the equivalent of one planeload every 8 minutes, 24 hours a day, for 9 years. At least 270 million cluster bomblets were dropped as part of the bombing campaign; up to 80 million failed to detonate.

Data from a new survey completed in Laos indicate that UXO, primarily cluster bombs, have killed or maimed as many as 50,000 civilians in Laos since 1964 (and 20,000 since 1974, after the war ended). Each year there continue to be close to 300 hundred new casualties in Laos. About 40% of accidents result in death, and 40% of the victims are children. Boys are particularly at risk.

From LEGACIES OF WAR
www.legaciesofwar.org

CULTURAL REMNANTS: THE HMONG

The Hmong village of Nasala.

Hmong families tie strings around wrists during New Year celebration.

Grandmother with tied bracelet strings.

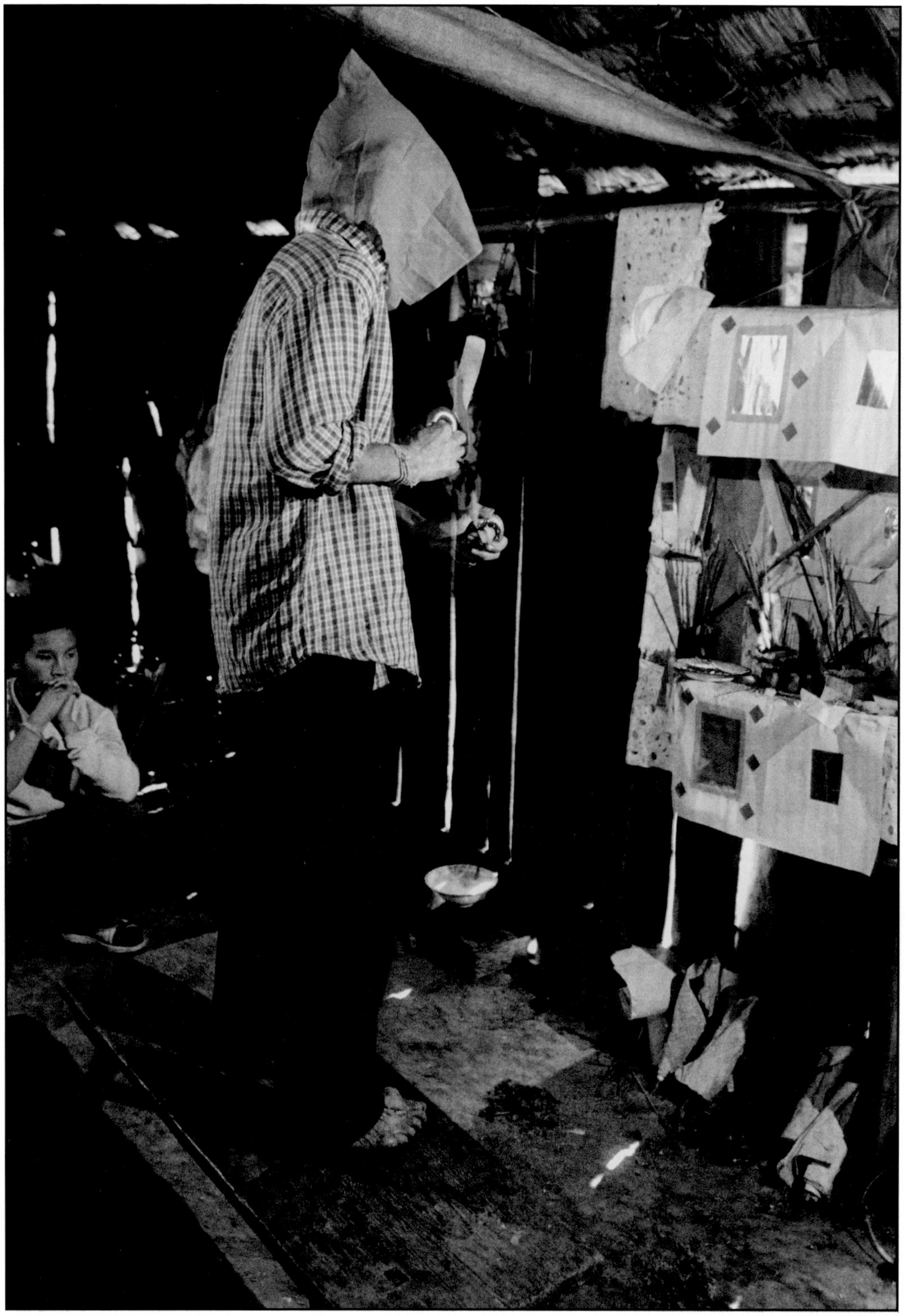

Shaman begins New Year ceremony at home.

Children watch grandmother get dressed.

Hair is burned on a cow for New Year feast.

Water buffalo fight during New Year.

Funeral.

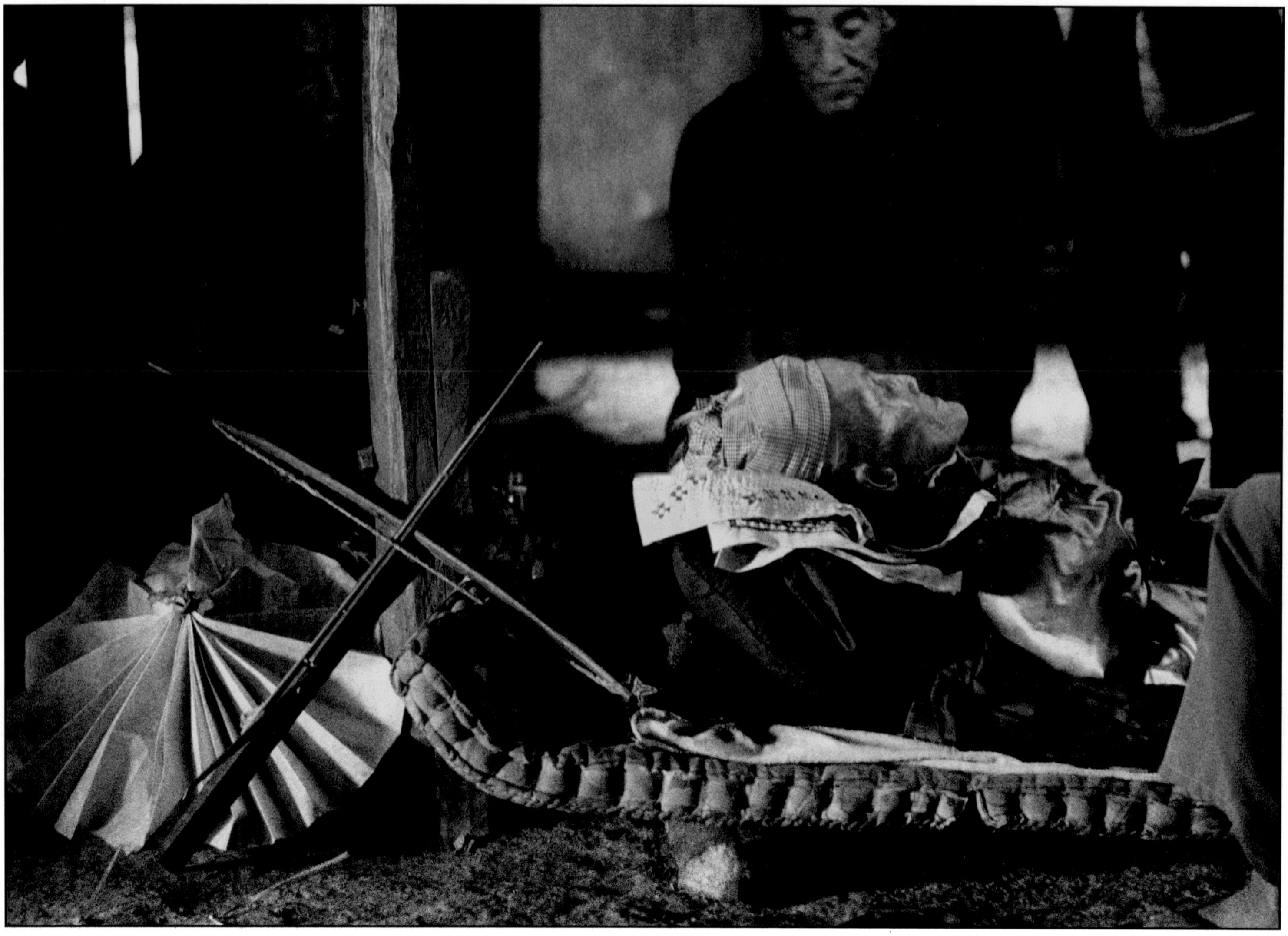

Funeral for elder, said to be over 100 years old.

Villagers celebrate end of Free Spirit ceremony.

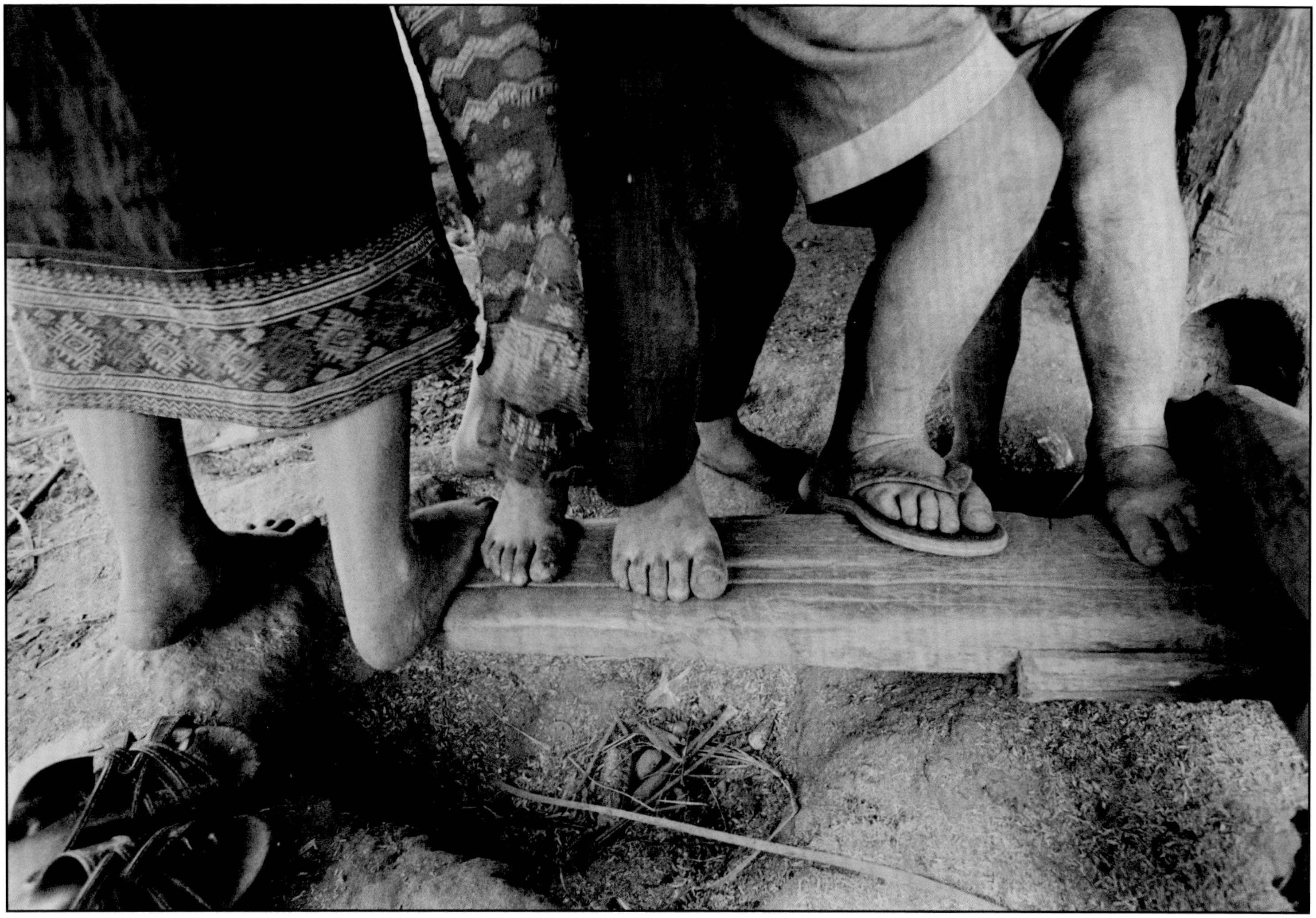

Children work a wooden mortar and pestle to pound rice.

Elder harvests rice.

Families work each other's rice fields.

Harvested rice is winnowed by mountain breeze.

Children care for younger siblings during growing season.

Children compete for sugarcane scraps.

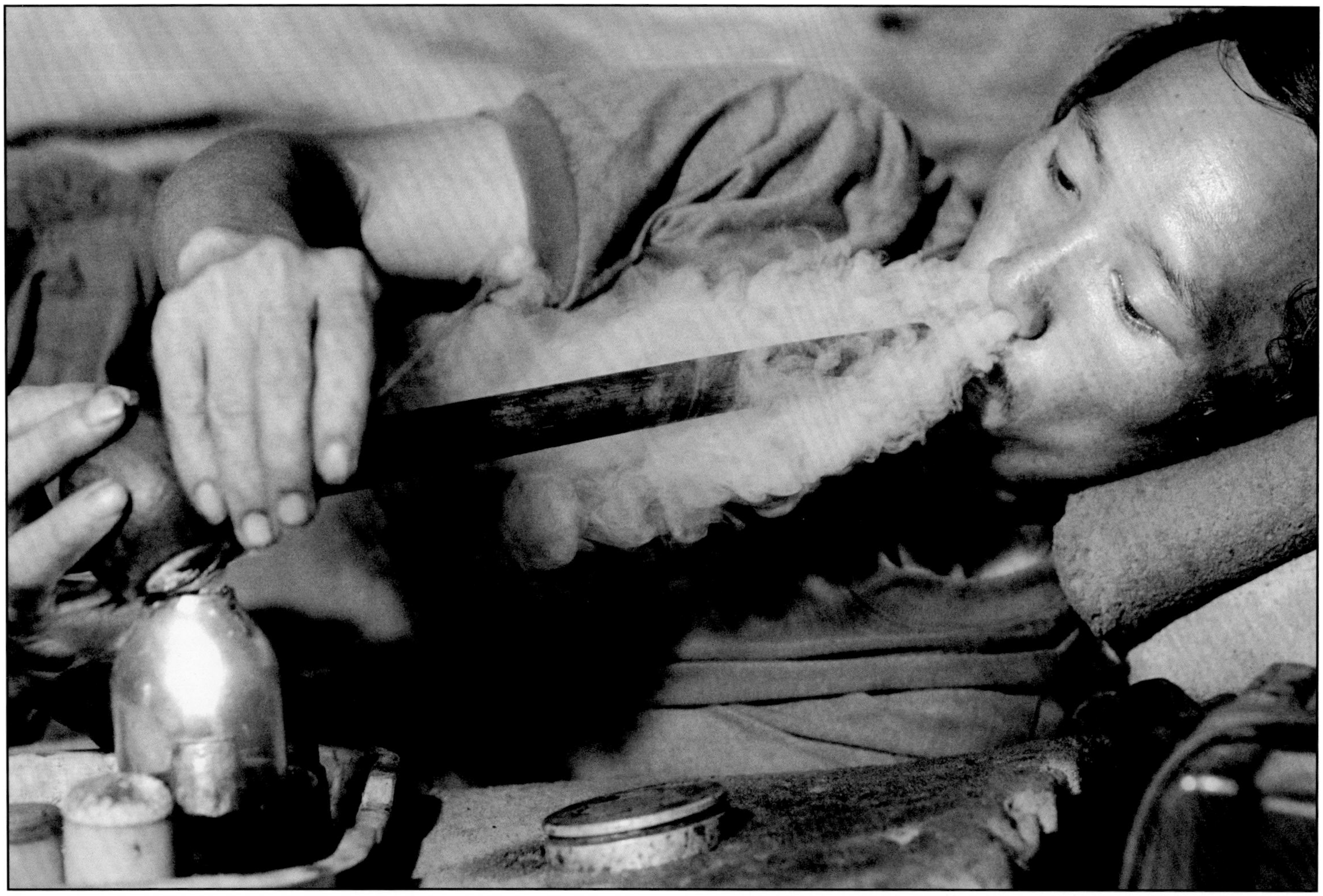

Smoking opium.

Family with newborn baby.

Mother and child attend funeral.

EXTENDED CAPTIONS

REMNANTS OF WAR

Pg 11: Defused bombs stand along Route 7 in Phonsavan, a reminder of the heavy bombing of the area during the Vietnam War.

Pg 12: A Lao soldier uses a metal detector as he looks for unexploded ordnance during a road-widening construction project along Route 7 outside of Phonsavan.

Pg 13: In a Phonsavan hospital recovery room, a Lao father holds a piece of shrapnel extracted from his son's arm after a cluster bomb exploded.

Pg 14: Hmong children walk past posters in a roadside market warning them not to touch unexploded ordnance, near the village of Nasala.

Pg 15: Hmong children gather around a discarded cluster bomb casing used as a fence in front of their home in Nasala.

Pg 16: A horse eats from a cluster bomb casing in Ban Tha Jok.

Pg 17: Spring onions grow inside a cluster bomb casing in Nasala.

Pg 18: Children peer through cluster bomb casings used as a garden fence in Ban Tha Chok.

Pg 19: Lao rice farmer Pho Vieng, shown here with his daughter, lost an arm and a leg when a cluster bomb exploded. After his injury he considered suicide but remembered he has two children to care for.

Pg 20: Xieng La, a Lao who fought with General Vang Pao and the Hmong army during the war, reminisces while holding a drawing depicting the fighting. The hand-drawn image was from a popular Lao currency note printed after the war. He supplements the income he earns as a rice farmer by defusing old bombs.

Pg 21: Xieng La taps a live mortar shell in an attempt to loosen and then extract gunpowder to sell. Local tourist guesthouse owners buy his defused bombs for decorations and souvenirs.

Pg 22: Blacksmith Con Toua Yang uses an artillery shell as an anvil while forging bomb shrapnel into a grass rake for rice farming in Nasala.

Pg 23: Rice farm laborers use grass rakes forged from defused bomb material to till the ground in Ban Tha Jok.

Pg 24: Blacksmith Papo Yang washes his hands after forging metal grass rakes from artillery shells in Ban Tha Chok.

Pg 25: In the Lao village of Napia, a man melts 41mm grenade shell casings and then pours the molten liquid into wooden casts to make the aluminum spoons. The spoons are used in nearby noodle shops. With the help of his family, he makes 500 spoons each weekday in the dry season.

Pg 26: A former bomb defuser holds two Chinese mortar rockets surrounded by defused U.S. bombs from the Vietnam War. He stopped the dangerous work after a neighbor was seriously injured.

Pg 27: A Lao scrap yard worker, at center, greets her children at the end of the day in Phonsavan.

Pg 28: A Lao soldier holds a newly unearthed live cluster bomb along Route 7. Over 260 million cluster bombs were dropped on Laos during the Vietnam War.

CULTURAL REMNANTS: THE HMONG

Pg 31: The village of Nasala in the northeast mountainous region in Laos is one of the Hmong resettlement villages in Xieng Khoung province.

Pg 32: Families tie sacred string bracelets around each other's wrists as part of the New Year festival in Ban Tha Jok. The strings are meant to preserve the soul and to ward off evil spirits.

Pg 33: Grandmother Le Vang is shown with wrist strings meant to preserve the soul and to ward off evil spirits. The strings were given to her by relatives during the New Year celebration in Ban Tha Jok.

Pg 34: A shaman begins the New Year ceremony in Nasala before a prayer altar covered with gold leafed paper.

Pg 35: Children, in Ban Tha Jok, watch their grandmother Le Vang get dressed.

Pg 36: Villagers burn the hair off of a slaughtered cow before a feast commemorating the New Year in Ban Tha Jok.

Pg 37: Farmers from nearby villages bring their water buffalo to compete in the annual New Year tradition in Nasala.

Pg 38: A funeral in Ban Tha Jok. Gilded paper is burned while a Hmong traditional flute (the qeej) is played.

Pg 39: Hmong villagers mourn a woman believed to have been over 100 years old. The wooden cross and bow represent a weapon to be used in the afterlife.

Pg 40–41: Villagers from Ban Tha Jok conclude the three-day Free Spirit Ceremony. A man spins a rice sifter then drinks rice whiskey from a bamboo shoot.

Pg 42: Children from Ban Tha Jok village use their feet to power a large wooden mortar and pestle that pounds the rice.

Pg 43: An elder harvests rice outside the village of Nasala.

Pg 44: Families work each other's rice fields in the annual harvest in November.

Pg 45: Freshly harvested rice is winnowed by a mountain breeze. In the northeast mountainous region, there is one rice harvest each year.

Pg 46: With parents tending to the rice fields, children care for younger siblings during the day.

Pg 47: Children in Ban Tha Jok fight over sweet sugarcane scraps after the juice is extracted at harvest time from a wooden foot-powered crusher.

Pg 48: Opium can be a cash crop for the Hmong during the dry season. Some villagers, like this man, get addicted. This man smokes opium, he says, to alleviate the pain from a bomb injury.

Pg 49: A mother cradles a newborn baby while the father smokes strong tobacco.

Pg 50: A Hmong mother and child attend a village funeral in Ban Tha Jok.

WRITING ON REMNANTS

When working in the field, Michael Greenlar resembles a chameleon adapting to its environment. Wearing black pants, a vest and a forest green shirt, he recedes into the background, often standing to the side to watch his subjects as they naturally interact with one another, creating various compositions he will capture. His photo essay, Remnants of a Secret War, is the result of 10 trips to a remote mountainous region of Laos to document the lives of the Hmong people, who have been forgotten by those responsible for their present circumstances.

Greenlar travels through seven airports under the guise of a tourist in order to reach Xieng Khoung Province. Every return trip, he brings along the prints from the last visit and a Polaroid camera to share his pictures with his subjects. Though he's admittedly drawn to people without a voice, the inspiration for Greenlar's work is often "just something that happens. Each trip you come away with a different part of the story." This story helps us imagine the physical reality of the Hmong people's lives in a land dotted with the remains of over 2 million tons of bombs dropped throughout Laos during the war in Vietnam. It also tells of the resourcefulness of a people who rely on each other, and on those very bombs, for their livelihood.

The Hmong are migrants who have always used natural resources as a means of survival. They are organized into clans, and each household is an active participant in the cultivation of crops. The children live constantly in the presence of adults, and gain knowledge about all aspects of their culture at an early age. The Hmong are careful to teach their beliefs in shamanism and animism, as well as their customs and survival methods, to each new generation. The ingenuity and resilience of these people enables them to reuse the deadly artifacts they've inherited as a critical source of metal, despite their extreme danger. Greenlar speaks vehemently about the bombs, particularly cluster bombs: "They have been using them in Iraq, Afghanistan, (and) Kosovo. They're really nasty little things. They're everywhere. So these are kind of like little calling cards that we have left all over the world. Cluster bombs are not considered land mines. They fall between the cracks in a lot of treaties and weapons bans."

His photographs juxtapose images of happiness and horror, grittiness and innocence, poverty and family. Scenes of blacksmiths converting the bombshells into tools for farming and into large pots for feeding their animals or planting flowers show how strong the Hmong are, and their incredible will to move forward. Their use of these shells as fences or feeding troughs represents their yearning and struggle for normalcy. The emotions on their faces, and the emotions we feel when we learn of their story, connect us as humans. Like so many people in our war-torn world, the Hmong ask for no more than survival.

Greenlar's use of black-and-white film emphasizes the scarcity in the scenes, which magnifies our sense of the suffering endured by these people. Black and white is less intrusive and more direct, allowing the viewer to focus on the content and message behind Greenlar's work. The strong lines